Here Is the Southwestern Desert

Madeleine Dunphy

ILLUSTRATED BY

Anne Coe

HYPERION BOOKS FOR CHILDREN

NEW YORK

Other books by Madeleine Dunphy

HERE IS THE ARCTIC WINTER
HERE IS THE TROPICAL RAIN FOREST

The artwork for each picture is done in acrylic.
This book is set in 14- and 16-point Hiroshige Book

Text © 1995 by Madeleine Dunphy.
Illustrations © 1995 by Anne Coe.
Printed in Hong Kong.
For more information address
Hyperion Books for Children,
114 Fifth Avenue, New York, New York 10011.
FIRST EDITION
1 3 5 7 9 10 8 6 4 2
Library of Congress Cataloging-in-Publication Data

Dunphy, Madeleine.
Here is the southwestern desert/Madeleine Dunphy; illustrated
by Anne Coe—1st ed.
p. cm.
ISBN 0-7868-0049-6 (trade)—ISBN 0-7868-2038-1 (lib. bdg.)
1. Desert ecology—Sonoran Desert—Juvenile literature.
2. Natural history—Sonoran Desert—Juvenile literature.   3. Sonoran
Desert—Juvenile literature.   [1. Desert ecology—Sonoran Desert.
2. Sonoran Desert.   3. Ecology—Sonoran Desert.]   I. Coe, Anne,
ill. II. Title.
QH104.5.S58D86 1995
574.5'2652'0979—dc20   94-9375   CIP   AC

For Deirdre
—*M.D.*

To my mother, Mary,

who taught me to love the desert
—*A. C.*

*H*ere is the southwestern desert.

*H*ere is the cactus

that is covered with spines

and can live without rain

for a very long time:

Here is the southwestern desert.

*H*ere is the hawk

that perches on the cactus

that is covered with spines

and can live without rain

for a very long time:

Here is the southwestern desert.

*H*ere is the lizard

who is spied by the hawk

that perches on the cactus

that is covered with spines

and can live without rain

for a very long time:

Here is the southwestern desert.

Here is the roadrunner

that chases the lizard

who is spied by the hawk

that perches on the cactus

that is covered with spines

and can live without rain

for a very long time:

Here is the southwestern desert.

*H*ere is the tree,

which shelters the roadrunner

that chases the lizard

who is spied by the hawk

that perches on the cactus

that is covered with spines

and can live without rain

for a very long time:

Here is the southwestern desert.

*H*ere is the sun

that blazes on the tree,

which shelters the roadrunner

that chases the lizard

who is spied by the hawk

that perches on the cactus

that is covered with spines

and can live without rain

for a very long time:

Here is the southwestern desert.

*H*ere is the bobcat

who basks in the sun

that blazes on the tree,

which shelters the roadrunner

that chases the lizard

who is spied by the hawk

that perches on the cactus

that is covered with spines

and can live without rain

for a very long time:

Here is the southwestern desert.

Here is the badger

that smells the bobcat

who basks in the sun

that blazes on the tree,

which shelters the roadrunner

that chases the lizard

who is spied by the hawk

that perches on the cactus

that is covered with spines

and can live without rain

for a very long time:

Here is the southwestern desert.

Here is the squirrel

who is unearthed by the badger

that smells the bobcat

who basks in the sun

that blazes on the tree,

which shelters the roadrunner

that chases the lizard

who is spied by the hawk

that perches on the cactus

that is covered with spines

and can live without rain

for a very long time:

Here is the southwestern desert.

*H*ere is the coyote

that pounces on the squirrel

who is unearthed by the badger

that smells the bobcat

who basks in the sun

that blazes on the tree,

which shelters the roadrunner

that chases the lizard

who is spied by the hawk

that perches on the cactus

that is covered with spines

and can live without rain

for a very long time:

Here is the southwestern desert.

*H*ere is the snake

that hisses at the coyote

that pounces on the squirrel

who is unearthed by the badger

that smells the bobcat

who basks in the sun

that blazes on the tree,

which shelters the roadrunner

that chases the lizard

who is spied by the hawk

that perches on the cactus

that is covered with spines

and can live without rain

for a very long time:

Here is the southwestern desert.

*H*ere is the hare

who hears the snake

that hisses at the coyote

that pounces on the squirrel

who is unearthed by the badger

that smells the bobcat

who basks in the sun

that blazes on the tree,

which shelters the roadrunner

that chases the lizard

who is spied by the hawk

that perches on the cactus

that is covered with spines

and can live without rain

for a very long time:

Here is the southwestern desert.

*H*ere is the cactus

that is food for the hare

who hears the snake

that hisses at the coyote

that pounces on the squirrel

who is unearthed by the badger

that smells the bobcat

who basks in the sun

that blazes on the tree,

which shelters the roadrunner

that chases the lizard

who is spied by the hawk

that perches on the cactus

that is covered with spines

and can live without rain

for a very long time:

Here is the southwestern desert.

The animals shown below live in the Sonoran Desert, which is located in parts of Arizona, California, and Mexico. There are more animal and plant species living in the Sonoran Desert than in any other desert in North America. Deserts are known for their clear skies, high temperatures, and lack of rain.

COLLARED PECCARY

CACTUS WREN

CHUCKWALLA

RED-TAILED HAWK

BLACK-TAILED
JACKRABBIT

KIT FOX

RINGTAIL

COYOTE

GREATER ROADRUNNER

COLLARED LIZARD

BADGER

BOBCAT

GOPHER SNAKE

A special note about two of the animals:
Although black-tailed jackrabbit is the proper name of the species pictured in this book, it is not really a rabbit but a hare. Rabbits are born naked and blind, while hares are born with fur and a well-developed sense of sight.

The collared peccary lives in many different environments—from the rain forests of Central and South America to the southwestern deserts of North America. Although its proper name is the collared peccary, it is more commonly known as the javelina.

DESERT
TORTOISE

Like many natural environments, deserts are threatened by human activities. If you would like to find out ways to help protect the desert, you can write to The Randall Davey Audubon Center, P.O. Box 9314, Santa Fe, New Mexico 87504.

ROUND-TAILED
GROUND
SQUIRREL